# PLANT PROTECTION IN THE PACIFIC 4

## SEMISI PONE
### BSc, MSc (Hons)

I wish to express my gratitude to the following people who helped me with my research work and surveys in Tonga during the collection of the information for this book.

1. Dr Viliami Manu, Deputy Director, MAFFF, Tonga.
2. Meleane Vea, Librarian, MAFFF Research Division, Vaini, Tonga.
3. 'Emeline, Research Assistant, Research Division, MAFFF, Tonga.
4. Tevita, Eastern Districts Agriculture Office, MAFFF, Tonga.
5. Samuela, Eastern Districts Agriculture Office, MAFF, Tonga.

# CONTENT

# List of Tables.

# INTRODUCTION.

This book is about the current status of several important plant diseases in Tonga, which is shaping the future of Tonga's food security. One of the most important issues discussed in this book.

I visited the island of Tongatapu, Kingdom of Tonga, from 28 January to 25 February, 2015 to survey and take pictures of the crops and fruit trees suffering from diseases for the production of this book.

## Citrus Tristeza Virus

I suspected Citrus tristeza virus (CTV) has wiped out the sweet orange (*Citrus sinensis*) and mandarins (*Citrus reticulata*) in Tonga. Even the wild limes and rough lemons, once common, are hard to find. It is not clear how fast this destruction was, but as I remember, my grandparent's property at Mu'a (Tatakamotonga) was full of sweet oranges and mandarins in the 1970s. But I have not seen any sweet orange or mandarin for many years, even before moving to Auckland in 1996.

It is a worldwide phenomenon, sweet orange and mandarins are reported to have been wiped out by the millions in South Africa, South America, Spain and the United States since the 1930s (Wikipedia).

It was inevitable that the virus will be introduced to Tonga, accidentally, sometime during the 1970s as large numbers of citrus especially sweet orange and mandarin were planted or had regenerated naturally. Every household had some sweet orange and mandarin, trees in town, or at their farm.

The virus was probably introduced without proper quarantine procedures as Tongan farmers are well known for bringing plants and seeds from overseas without informing the local MAFF Quarantine staff.

Citrus Tristeza Virus is spread very effectively by *Aphis gossypii.* Aphid activity will determine the spread of the virus.

## Vanilla Necrosis Potyvirus

Vanilla Necrosis Potyvirus is probably still a 'low' threat to vanilla (*Vanilla fragrans*) in Tonga. After the initial research and control measures implemented (Pone, 1988). There was a huge reduction in virus incidence in farmer's fields since control measures were implemented in the late 1980s.

## Cucumber Mosaic Virus

Cucumber Mosaic Virus is a very serious disease of kava (Davis *et al*, 1996). It remains a threat to kava production in Tonga despite the intensive research and control measures implemented by ACIAR and MAFF, Tonga .

## Pawpaw disease

A foliar yellowing 'disease' was observed on pawpaw. It is not clear how widespread it is. There are stories that some pawpaw plantations were wiped out by a disease (Manu pers. comm). This 'disease' needs to be investigated further as a matter of urgency.

## Anthracnose of yams

Anthracnose is a common problem of many crops in Tonga. Yam dieback is probably the most important. Anthracnose normally attacks the early yams, mainly the variety 'kahokaho', which is the most popular and most cultivated variety.

It is another 'explosive' disease that will cause a lot of problems if it attacks other varieties of yams.

It also coincides with the anthracnose of mangoes and other tree crops which flower at the same time when the early yams are planted around June-August.

During 'wet years', anthracnose can cause a 100% loss of the mango crop in Tongatapu… and other treecrops like avocado. Tava may also be affected. Local growers have noticed this phenomenon and named it ''auhia'. They believe it is the rain that washed away the mango flowers and young fruits hence the name ''auhia' which means 'washed away' in the Tongan language.

The scientific literature and local MAFF research staff established that it is a fungus, *Colletotrichum gloeosporioides*, which cause the problem (Holo, pers. comm).

## Tree Species that are 'disappearing'.

Avocado was very common in Tonga 30-40 years ago. Almost every property in our suburb in Nuku'alofa had one or two trees. It is one fruit tree species that is disappearing from Tongatapu. Others include Vi (*Spondias*), fekika (mountain apple) and even some varieties of breadfruit. It is not clear why their numbers are declining.

*Phytophthora cinnamomi* is known to cause problems with avocado in other countries and could be the case in Tonga. Other

*Phytophthora* species that may be important include *P. palmivora* on breadfruit. Again this should also be investigated as a top priority.

Only mangoes and tava (*Pometia pinnata*) are still very common and even increasing in numbers in the last 1-2 decades.

Some comments will be made about the widespread problem of 'sooty mold' which is 'rampant' on Tongatapu.

The chapters of this book discuss the above diseases in light of the survey information with additional discussions and 'facts' from various regional and international sources.

National Plant Protection Organisations

This book is not intended as a textbook but 'sharing' of information with the National Plant Protection Staff (NPPS) in Tonga as well as around the Pacific. Most of the disease problems discussed are also present in neighboring countries.

It was the intention for the establishment of the Pacific Plant Protection Organisation (PPPO), in 1994, as one of the Regional Plant Protection Organisations (RPPOs) to allow the Pacific Island members to contribute and share in global Plant Protection information available from FAO and other RPPOs. This would enhance trade through better Plant Quarantine Standards and practices for all.

# CHAPTER 1. CITRUS TRISTEZA VIRUS

The loss of 'hundreds of thousands' of sweet orange and mandarin trees from Tongatapu and the islands of Tonga, during the 1970s, was a great food security threat to the Kingdom. Citrus Tristeza Virus*, and probably psorosis virus* wiped out these important fruit trees within a decade. Later generations of Tongans missed out on the 'bumper crops' when sweet orange and mandarin trees 'bend over' with the weight of the ripe fruits all over the Kingdom. It is the saddest moment for many, including myself, to remember those days of the 'abundance of sweet orange and mandarins' in the Kingdom, which are now gone forever.

\* - Both these viruses are recorded by Mossop and Fry (1984) as present in Citrus in Tonga.

Citrus tristeza virus was recorded on *Citrus sinensis* or sweet orange and Citrus psorosis on *Citrus limon* in Tonga by D.W. Mossop and P.R. Fry, DSIR, New Zealand in 1984.

That was the last major survey of plant diseases in Tonga sponsored by SPEC, UNDP and FAO.

However, only Citrus Tristeza Virus is recorded in the literature as the sole 'destroyer' of millions of sweet orange and mandarin in many countries (Wikipedia).

Since the 1960s in Tonga, there was an abundance of sweet orange and mandarins on Tongatapu. I can remember almost every house and farm, I visited as a child, had several trees. My grandparent's house at Mu'a had several sweet orange and mandarins. They grew in abundance, not only in my grandparent's backyard, but also in the surrounding bush and farmland.

During the season, fruits on sweet orange and mandarin trees were so numerous that branches bend under their weight. It was a delight for us children to climb and pick the fruits or just stand on the ground and pick the fruits from the lower branches.

Now you cannot find a sweet orange (moli kai) or mandarin (moli peli) plant on Tongatapu. My short survey in February 2015 confirms it. Although I have not done any surveys in the outer islands, I have been told by locals, on Tongatapu, that they have also disappeared from the islands of 'Eua, Ha'apai, Vava'u and the Niuas, as well.

There are 2 local varieties known as the 'moli vaikeli' which is similar to 'moli peli' but the fruits are only about a third of the size of a moli peli. I will refer to them as the 'lesser' mandarin.

The loss of sweet orange and madarin is a huge loss to Tonga because they were such perfect, sweet, juicy and productive species loved by all the kids and villagers. Excellent sources of vitamin c.

## Looking for the sweet orange and mandarin...

Part of the survey was simply walking around Nuku'alofa for several days and taking note of the citrus plants and other trees in people's front yards (see Notes on Survey). Pictures were taken to record the trees present.

Although there were limes and 'moli uku' no other citrus species can be seen. Rough lemon and limes were very common in Nuku'alofa, they are now not as abundant.

During my time working as a Plant Pathologist/Senior Plant Virologist for MAFF, Tonga, in 1985-1992, I had already noticed the tristeza 'wilt' of citrus. I had witnessed the yellowing, wilt and defoliation of several citrus trees. The noticeable symptoms of the disease.
I estimate that during the later part of the 1970s, most of the moli kai and moli peli of Tongatapu had 'disappeared' or 'wilted' and died due to citrus tristeza virus (CTV) infection.

Other citrus species like the rough lemon (lemani) and pomelo ( moli tonga) are also disappearing (see Table 3).

Interestingly, 'moli vaikeli' seem to be the most common especially in the eastern districts of Tongatapu.

In the New Zealand supermarkets the 'moli peli' is referred to by several names including mandarin and tangelo. It is also referred to in the FAO ECOCROP website as tangerine and satsuma orange. A New Zealand mandarin appear very similar to one of the 2 'moli vaikeli' varieties in Tonga (see Figure 5 & 6).

The mandarin known as 'moli peli' can be seen in Figures 3 & 4. Interestingly, the mandarin in Figure 3 is very similar in appearance to one of the 'moli vaikeli' varieties in Tonga, but they have smaller fruits.

Sweet orange too is also known by several names including New Zealand Orange, Navel Orange, USA Orange, Australian Orange and Valencia Orange (see Figure 7, 8 & 9).

Sweet orange varieties were planted, experimentally, at the MAFF Vaini Research Station in Tonga. Although they look exactly the same as the 'moli kai/moli inu' that was widely grown in Tonga, they may be different varieties. They were grafted, with other citrus such as lime, and sold to the public in the 1980s. It would be interesting to try and find out what happened to those trees.

Moli kai, in Tongan, means 'orange for eating'. It is also referred to as 'Moli inu' or 'orange for drinking'. Moli peli probably refers to the skin of the mandarin which seem to have lots of 'pits' on it. Moli vaikeli means 'orange near the water well'. Maybe the first moli vaikeli was found near a waterhole! Moli uku is 'orange for washing the hair', Moli Tonga is ' Tongan orange'.

Figure 1. This is what the 'moli uku' fruits look like. They are the same size and look like other oranges but have a sour taste like a cross between a grapefruit and a lemon. This tree was found on the roadside at Kolovai village, Tongatapu (Figure 2).

Figure 2. A 'moli uku' plant bearing a large number of fruits that look exactly like the 'moli kai' (sweet orange). This tree was found on the roadside at the village of Kolovai. The property owner said this orange tree is very sour. It is probably sour or bitter orange (*Citrus aurantarium*).

# SOME PICTURES OF SWEET ORANGE AND MANDARIN.

No sweet orange or mandarins can be found in Tonga, so these pictures were taken in two Auckland supermarkets. Mandarin was labelled as tangelo in one supermarket. It is also recorded in Mossop and Fry (1984) as tangerine.

Figure 3. United States Mandarin at a supermarket in Auckland, New Zealand. These fruits are exactly the same as the Tongan 'moli peli' which was wiped out by tristeza virus. The moli peli, in some cases, were much bigger than these fruits.

Figure 4. Mandarin is also referred to as Tangelo in another Auckland supermarket. They may be different varieties but they look almost the same. As in the case of 'moli peli' in Tonga, all the fruits that look similar were put under this name. Moli peli on Tongatapu has been wiped out by tristeza virus.

Figure 5. This species is also referred to as New Zealand Mandarin in this supermarket. The fruits are slightly smaller than Mandarins from the United States and the Tangelo. This fruit is very similar to 'moli vaikeli' in Tonga.

Figure 6. New Zealand Mandarin is referred to as 'moli vaikeli' in Tonga. This variety is still common on Tongatapu. There are two distinct varieties referred to as 'moli vaikeli', probably natural hybrids.

Figure 7. This is sweet orange from the United States in an Auckland supermarket. This fruit is very similar to 'moli kai' also known as 'moli inu' in Tonga which has been wiped out by the tristeza virus.

Figure 8. This is the New Zealand sweet orange which is slightly different from the United States variety. It is also very similar to the 'moli kai/moli-inu' in Tonga.

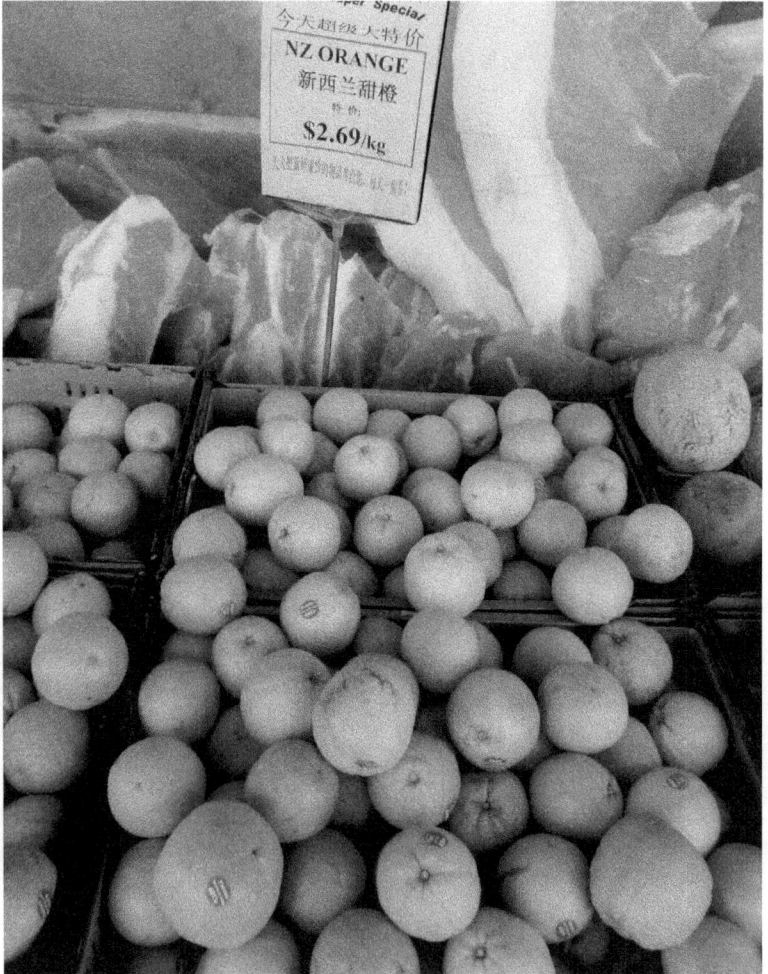

Figure 9. These are Australian navel sweet oranges are also very similar to 'moli kai/moli-inu' in Tonga.

Figure 10. This New Zealand lime in an Auckland supermarket is very similar to limes in Tonga, which are still common on Tongatapu.

New Zealand limes can be grown in Tonga. Tongan growers should buy some potted limes in New Zealand to try out at home. All they need is a MAFFF Quarantine permit from Tonga. They should also enquire about any requirements in New Zealand. Tip. Ask the 'Nursery Supplier' if they have any pot plants resistant to tristeza and psorosis viruses.

The survey from January 29, 2015 to 12 February 2015 was to assess the situation with the sweet orange or 'moli kai' (*Citrus sinensis*) and mandarin or 'moli peli' (*Citrus reticulata*). The survey did not find any sweet orange or mandarin, although I came across 2 'moli uku' plants, several 'moli vaikeli' and several lime trees.

I have noticed the disease during my time with MAFF, Tonga that citrus species were in decline.

So what is the problem with the disappearance of these citrus fruits from Tongatapu? Or even the whole island Kingdom of Tonga?. We'll, probably none, as most people don't even notice they have disappeared!. But they were excellent sources of vitamin c for adults and village children, all over Tonga. When they were in season, the fruits were so numerous that each villager would probably eat several oranges or mandarins everyday. They were sold in the main market, sports tournaments and shows and even by the roadside on

Tongatapu. Now villagers have less choices of 'cash' fruits and sources of vitamin c.

There was a 'tree crop project' at the Vaini Research Station, MAFF, Tonga for many years during the 1980s, 1990s. It did not appear to have any impact on local fruit production. Popular citrus varieties were grafted and sold to the public, but there is very little evidence of that, in the field, during this survey.

Tree crop project...

There need to be an ongoing tree crop project. Grafting popular citrus species like oranges, mandarin, limes and lemons to sell to the public.

Popular varieties of mangoes, tava, vi, fekika, lesi, breadfruit and so on should be supplied in pots ready for planting, as well.

Coconuts are also getting old and need to be replanted. There are some large areas on the Eastern District without any coconuts.

Figure 11. One of the very few remaining rough lemons, we came across during this survey, growing wild in a cattle farm paddock at the village of Kolovai, Western District, Tongatapu.

History of citrus fruits in Tonga...

It is not clear how the several species of citrus were introduced to Tonga. Probably by the early missionaries and copra producers in the 19th century.

Figure 12. This is one of the two trees, I found, that looks like sweet orange or 'moli kai' (*Citrus sinensis*). But as in the case of the tree found on the road side at Kolovai. This tree could be a variety of local citrus known as 'moli uku' (*C. aurantium*) which looks exactly the same as 'moli kai'. Its juice taste like a cross between grapefruit and lemon. This tree was found in the Nuku'alofa suburb of Longolongo during my 'survey walk' in a fenced yard. Note the 'chlorosis' or yellowing symptom (bottom left corner) on some of the leaves. It may be the beginning of the 'tristeza wilt'.

Plant Virus Survey by Mossop and Fry
(1984). (Funded by FAO, SPEC, UNDP).

Citrus Psorosis Virus is present in Samoa
and Tonga. Citrus Tristeza Virus is recorded
in Cook Is, Fiji, Niue, Samoa, Tonga.

# Citrus tristeza virus (CTV).

CTV is a viral species of the *Closterovirus*
genus that causes the most economically
damaging disease to citrus. The disease has
led to the death of millions of *Citrus* trees
all over the world and has rendered other
millions useless for production. Farmers in
Brazil and other South American countries
gave it the name "tristeza", meaning sadness
in Portuguese and Spanish, referring to the
devastation produced by the disease in the
1930s. The virus is transmitted by aphids. In
Tonga, *Aphis gossypii* maybe the most
important vector.

CTV is a flexuous rod virus with dimensions of 2000 nm long and 12 nm in diameter.

In Tonga, sweet orange (*Citrus sinensis*) and mandarin (*Citrus reticulata*) were the most important species affected by CTV. Both species have now disappeared from Tongan farms and villages.

CTV is distributed worldwide and can be found wherever citrus trees grow.

Symptoms of CTV infection are highly variable and depend on several factors including host, virulence of the particular virus strain and environmental conditions. The three most common symptoms are decline (quick and slow), stem-pitting, and seedling yellows.

Decline is generally exhibited with sweet orange, mandarin, The decline may also be quick, resulting in host death just days after the first symptoms are noticed.

Stem-pitting is another symptom of CTV that manifests in most host types. The host will develop pits in the trunk and stem. This results is decreased tree vigor and reduced fruit yield. This is typically caused by the more virulent strains of CTV.

The third major symptom of CTV infection is seedling yellows. Symptoms include yellowing of foliage and general dieback.

CTV is classically diagnosed with highly predictable symptoms on the leaves begin as clear veins that turn corky, which is then followed by chlorosis and cupping of the leaf.

Electron microscopy and ELISA can be used to test for CTV.

CTV is a virus that is limited to the phloem tissues of its host. It is transmitted semi-persistently by vectors that penetrate the phloem to extract sap, mostly the aphid species that colonize the crop. The brown citrus aphid is considerably more efficient at transmitting the virus than are other aphids that infest citrus. In Florida, it has been shown to be from six to twenty five times as efficient as *Aphis gossypii*, the most efficient vector found in the state before the introduction of the brown citrus aphid.

This efficiency is enhanced by the narrow host range of the brown citrus aphid and its tendency to produce winged forms in order to colonize new growth. _A. gossypii_ has a much wider host range, including hundreds of plant species in Florida. _Aphis gossypii_ (melon and cotton aphid) is the most common aphid in Tonga.

Aphids are the main vector by which CTV is transmitted. In the United States _A. gossypii_ require at least 30 to 60 minutes of feeding to acquire the virus, and remain viruliferous for at least 24 hours after. _Aphis gossypii_ has a 78% transmission efficiency.

When CTV was first discovered quarantine was the best management strategy, now quarantining only works for areas where a small amount of trees are infected. If it is possible to keep the field permanently free of CTV, the planting of virus-free trees is also practical.

CTV is the most economically important and damaging virus of citrus trees. It can be spread quickly and do damage not only by killing trees but also by stem pitting normal citrus trees. It has killed more than 80 million trees worldwide, mainly in South Africa since 1910, Argentina (10 million) and Brazil (6 million) since 1970, and the U.S. (3 million) since 1950. In Spain there has been a progressive decline in production from over 40 million sweet orange and mandarin trees (Wikipedia).

---

In Tonga, decline of sweet orange and mandarin production in the 1970s has resulted in the complete disappearance of these important fruit trees in the 1990s. This survey confirms it. No sweet orange or mandarin tree were found or seen in the survey area (See Survey Map).

Tonga needs a lot of orange and mandarin trees...

There are no orange and mandarin trees left on Tongatapu. There is a local species of orange, similar to 'moli peli', called 'moli vaikeli' but it has small fruits and is not popular. An active replanting of both sweet orange and mandarin by growers themselves, with potted plants from New Zealand, is the best and quickest way to solve this problem.

New Zealand, especially, Auckland and Northland have similar weather and temperatures to Tonga. There are a large number of citrus species available from nurseries in these areas which can be planted in Tonga. Tongan visitors to New Zealand can simply buy them and take them home to try. Note; A permit from MAFFF Quarantine in Tonga will be required for introducing plants from overseas into the country.

## POSSIBLE SOLUTION TO THE CITRUS FRUIT DECLINE ON TONGATAPU.

The dwarf orange trees shown in Figure 13 and 14 appear to be mandarin happily growing in a garden in Auckland. The fruits are larger than the 'moli vaikeli' in Tonga but smaller than the 'moli peli'. Its size and early fruiting ability makes it perfect for the front yards of properties in Nuku'alofa and villages throughout the Kingdom.

Figure 13. A dwarf early fruiting mandarin in a front yard garden in Auckland, New Zealand.

Figure 14. A very productive dwarf, ripening mandarin in a home garden in Auckland, New Zealand.

The ripening fruits will be ready for harvest in a few weeks. These pictures were taken in March, 2015. These trees appear to be 'high production' varieties. A small sized tree will provide a large number of fruits for the family and relatives.

Similarly, early production lemons and limes are also seen in gardens around

Auckland, which has similar weather to Tonga. They can also be introduced as a solution to the 'loss of citrus trees' in Tonga.

---

# RECOMMENDATION.

There are a large number of Tongans travelling to Auckland every year. It is highly recommended that they apply for a permit from MAFFF Quarantine in Tonga, to bring dwarf oranges, mandarins, lemons and lime trees to Tonga for planting on their properties. Once in Auckland, they can buy these excellent early fruiting varieties and take them home. If they are planted six months to a year before next flowering season, they will bear fruit within a year!. The MAFFF should allow 20,000 oranges and mandarins, 10,000 each for lemons and limes, or similar numbers to satisfy the local demand. It is a quick and easy solution. <u>TIP: Ask the supplier of the dwarf trees if they can get you tristeza and psorosis virus resistant varieties that will grow well in Tonga.</u>

---

# CHAPTER 2. A NEW PAWPAW 'DISEASE'?

Pawpaw (*Carica papaya*) virus symptoms look similar to papaya lethal yellowing virus. The crown turns yellow and there is progressive 'dieback' of the tree from the crown. Symptoms observed on pawpaw appear to be virus induced. Yellowing of the crown, stunted growth and general dieback of the whole plant are typical virus induced symptoms.

Figure 15. This is one pawpaw plant that definitely shows 'viruslike' symptoms of leaf yellowing and stunting of new leaves. Suburb of Halaleva, Nuku'alofa, Tongatapu.

Although this 'disease' cannot be confirmed at this time. It is important to avoid spreading it through seed and vectors. Infected plants should be destroyed as soon as they are discovered by chopping the tree and burning or burying in the ground.

Figure 16. The pawpaw plant on the right is stunted with yellowing of the leaves. This is another definite 'viruslike' symptom. Note the vigorous growth of the plant on the left which was planted at the same time. The whole row of pawpaw are about the same size except the stunted plant. Eastern District, Tongatapu.

'Lethal yellowing' of pawpaw is one 'symptom' to be on the look out for in Tonga. There is a 'conspicuous' absence of pawpaw fruits from the Talamahu Market or roadside markets during my visit. It could be the effect of the decline in overall number of pawpaw plants and production of fruits in Tongatapu.

---

'Disease' control

Although the identity of this pawpaw 'disease' has not been confirmed, it is worth controlling its spread by doing some simple control measures.

Pawpaw plants showing disease symptoms of leaf yellowing and stunting should be removed immediately and destroyed by chopping up the plant then burning or burying underground. Use only seed from 'healthy looking' pawpaw trees for replanting.

---

Once there were pawpaw plants everywhere. Every household and farm had several trees. Now it is hard to find a pawpaw tree in Nuku'alofa. Like citrus, there is continuous regrowth from the seed bank in the ground.

I have taken pictures of, at least 2 pawpaw trees, showing symptoms (Figure 15 & 16). The crown of the plant is stunted, turns yellow or chlorotic then dies back. The whole process can occur very quickly and usually the plant collapses on the ground and disappears.

According to Dr Viliami Manu (pers. comm.), Deputy Director, MAFFF, Tonga there was a crop of experimental pawpaw planted at the Vaini Research Station some years ago. All the plants suddenly exhibited leaf yellowing and stunting, then they all wilted and die. The 'cause' is still unknown.

Although pawpaw plants can still be found in Nuku'alofa and all the villages of Tongatapu. They are definitely less in numbers now than 3 decades ago. Once there were one to several trees in almost every home garden in Nuku'alofa, now only one or two can be found among several home gardens, if at all. I noticed during my 'survey walk' around Nuku'alofa that they were not as common as they were when I was in High School more than 3 decades ago. This important fruit will probably also disappear within a few years if no disease control measures are put in place. Some farmers appear to be planting large numbers of pawpaw (See Figure 16). It is a good sign.

## Symptoms and Signs of PRSV.

*Papaya ringspot virus* infects papaya and cucurbits systemically is a potyvirus.   Symptoms on papaya are somewhat similar to those on cucurbits. In papaya, leaves develop prominent mosaic and chlorosis on the leaf lamina, and water soaked oily streaks on the petioles and upper part of the trunk.   Severe symptoms often include a distortion of young leaves which also result in the development of a shoestring appearance that resembles mite damage.   Trees that are infected at a young stage remain stunted and will not produce an economical crop.

www.apsnet.org

# CHAPTER 3.    VANILLA NECROSIS POTYVIRUS (VNPV).

Five vanilla (*Vanilla fragrans*) plantations in the Eastern Districts were inspected for VNPV symptoms but none were found. The recent 4-5 month drought from August 2014 to January 2015 has affected the plants so much that most look weakened with a 'grayish color' on the older stressed leaves.

During the survey it was obvious that the recent rain, in the previous 2 weeks, fed fast growing weeds which covered the once bare ground.

Pone (1988) had described the symptoms of VNPV and disease pathogenicity, serology and epidemiology. Pone and Pearson (1988) also made recommendations for controlling VNPV. Pone (2014) in his series of books 'Plant Protection in the Pacific' also discuss the VNPV symptoms and spread in some detail.

Figure 17. A 4-5 year old vanilla plantation, looking healthy although 'a bit stressed' from the long drought. Eastern District, Tongatapu.

Five plantations were inspected, on this trip, with no mottling, necrotic lesions, chlorosis or symptoms of any kind observed. But some plants appear stressed with 'lighter green leaves' (Figure 17).

Figure 18 . Vanilla plants were 'recovering' in this 5 year old plantation, although dead vines can be seen in this picture. The previous 4-5 month drought maybe the cause.

# CHAPTER 4. KAVA CUCUMBER MOSAIC VIRUS (CMV).

The Cucumber Mosaic Virus (CMV) problem of   kava (*Piper methysticum*) in Tonga is still common, judging from the number of infected plants found. Most of the plants observed during this short survey show symptoms of dieback. Chlorotic young leaves, necrosis of leaves and stem and even dead stems on some plants.

Figure 19. A 1-2 year old kava plot at Fua'amotu, Eastern Districts of Tongatapu. All the plants are dead with just the dry stems still standing due to the 4-5 month drought.

Crinkling of the new leaves and dieback of the new shoots are also very common symptoms of dieback.

Figure 20. These young 6-12 month old kava plants survived the 4-5 month drought although they look a bit stressed. The plants look healthy with no CMV symptoms. CMV presence in young kava plants may be the difference between survival or death in a drought. This plot was adjacent to the one in Figure 10.

The presence of CMV in kava plants obviously weakens the plant. This is often observed on plants which are not growing vigorously. Leaves 'droop' and appear 'wilted'. Leaves are usually partly necrotic.

---

*Cucumber mosaic virus* infection of kava (*Piper methysticum*) and implications for cultural control of kava dieback disease. By R. I. Davis,   M. F. Lomavatu-Fong, , L. A. McMichaef,   T. K. Ruabete,   S. Kumar and U. Turaganivalu.

*Cucumber mosaic virus* (CMV) was found by reverse transcription polymerase chain reaction (RT-PCR) to be not fully systemic in naturally infected kava (*Piper methysticum*) plants in Fiji. Twenty-six of 48 samples (54%) from various tissues of three recently infected plants were CMV-positive compared with 7/51 samples (14%) from three long-term infections (plants affected by dieback for more than 1 year). The virus was also found to have a limited ability to move into newly formed stems. CMV was detected in only 2/23 samples taken from re-growth stems arising from known CMV infected/dieback affected plants. Mechanical inoculation experiments conducted in Fiji indicate that the known kava intercrop plants banana (*Musa* spp.), pineapple (*Ananas comosus*), peanut (*Arachis hypogaea*) and the common weed *Mikania micrantha* are potential hosts for a dieback-causing strain of CMV.

Australasian Plant Pathology (see References).

Figure 21. Another kava plot affected by the drought. All the plants are dead with just the dry stems still standing. This was a 1-2 year old kava plot in a mixed crop with vanilla. The recent 4-5 month drought is blamed as the cause of the death of all the kava in this plantation. CMV may have weakened the plants. Note the quick recovery of the weeds after 2 weeks of rain. This plot was found on the Eastern District near Fua'amotu, Tongatapu.

No mature kava plantations were found during our survey, most have been harvested due to the effect of the 4-5 month drought, according to one of the Agriculture staff in the Eastern Districts (Samuela pers. comm.).

Most of the older kava plants (2-3 yrs or older) observed were in 'gardens' around Government Offices at Vaini Research Station, MAFFF and the Tourist Offices at Fa'onelua Park, Nuku'alofa.

Note the following pictures from some of the older kava plants seen. They are obviously weakened by the 4-5 month drought. They show 'drooping' and partly necrotic leaves, dieback but are still producing new shoots, which appear healthy, at the base.
The photos in Figure 22-25 were taken at Fa'onelua Park near the Tourisim Office, Nuku'alofa, Tongatapu.

Figure 22. This kava plant is showing the effect of the drought as well as the CMV. Note the drooping petiole and chlorotic/necrotic leaves. These are sure signs that CMV is present. This plant was one of several found at the Tourism Office garden at Fa'onelua Park, Nuku'alofa, Tongatapu.

Figure 23. A kava plant at the Fa'onelua Park showing CMV symptoms. Expanding necrotic area on the leaves. The light colored leaves maybe due also to the recent 4-5 month drought. This plant has a darker colored stem and maybe a different variety from the plant in Figure 13.

Figure 24. One of the symptoms of CMV on kava is the 'expanding' necrosis on the leaves. Although they may appear vigorous it can collapse within a matter of days when the stem become necrotic and breaks, usually at the nodes.

Figure 25. This plant shows advanced CMV infection. The terminal shoot has died and regrowth can be observed. The leaves are light coloured with advancing necrotic areas.

Pictures 25 and 26 were taken at the
MAFFF Vaini Research Division garden.

Figure 26. The first symptoms of Cucumber Mosaic
Virus (CMV) on young kava leaves is the chlorosis
of the leaf and veins, turning necrotic at the edges.
Note the 'crinkling' or 'bubble like' symptoms on
the young leaf on the left. The beginning of a
necrotic lesion on the edge. It will expand and kill
off the leaf and terminal shoot starting the 'dieback'.
The lower stem collapses in the advanced stage of
the dieback. This 3-4 year old plant was found in
the MAFFF Research Station garden, Vaini,
Tongatapu.

Figure 27. Advanced CMV infected kava plant with necrotic and severe crinkling and distortion of leaves. The whole shoot usually turn necrotic and rot, starting from the top and the lower stem collapses when stem lesions become necrotic. (Note. *Collectotrichum* which causes 'anthracnose' of most trees   and crops in Tonga is almost always found on the kava plants at the early stages of the 'rot'. It has led to conclusions in the past that *Collectotrichum* maybe the cause of kava dieback. It is referred to as a 'secondary pathogen' in the case of kava dieback.

Although only 3 plantations were surveyed and about 8 plants were found, in other areas, during this survey, most of them show symptoms of CMV. It seems that the disease control strategies are not working and need to be reiterated.

| Table 1. CMV CONTROL STRATEGIES. |
|---|
| 1.  Only use cuttings from healthy, non-symptomatic plants.<br>2.  Planting in a mixed crop will reduce disease incidence if it is accidentally introduced from outside the plantation by aphids.<br>3.  Partial shade appear to be conducive to plant vigor.<br>4.  Remove all CMV hosts from the immediate vicinity of the kava plot.<br>5.  Check the plot every week and remove any plants showing CMV symptoms.<br>6.  Replant with healthy cuttings.<br><br>Davis et al (1996), Davis (1999), Davis (2005). |

# CHAPTER 5. BREADFRUIT DIEBACK.

The following is a description and history of the breadfruit from wikipedia, the free online encyclopedia.

Breadfruit (*Artocarpus altilis*) is a species of flowering tree in the mulberry family, Moraceae originating in the South Pacific and that was eventually spread to the rest of Oceania. British and French navigators introduced a few Polynesian seedless varieties to Caribbean Islands during the late 18th century and today it is grown in some 90 countries throughout South and Southeast Asia, the Pacific Ocean, the Caribbean, Central America and Africa.Its name is derived from the texture of the cooked moderately ripe fruit, which has a potato-like flavor, similar to freshly baked bread.

Ancestors of the Polynesians found the trees growing in the northwest New Guinea area around 3,500 years ago. They gave up the rice cultivation they had brought with them from Taiwan, and raised breadfruit wherever they went in the Pacific (except Easter Island and New Zealand, which are too cold). Their ancient eastern Indonesian cousins spread the plant west and north through insular and coastal Southeast Asia. It has, in historical times, also been widely planted in tropical regions elsewhere.

Breadfruit (*Artocarpus altilis*) has been an important staple in Tonga and the Pacific Islands since recorded history began. It was so popular in the Ha'apai Group that a Tongan proverb describe how the islanders crave for it every morning. *'A faka'amu mei'*, means 'craving for breadfruit in the morning' in the Tongan language.

Breadfruit is an excellent food source in the islands. The story of Captain William Bligh and the mutiny on the Bounty involved a lot of breadfruit. Captain W. Bligh was given the task of collecting and transporting breadfruit seedlings from French Polynesia for planting in the islands of the Caribbean for the slaves in the sugarcane plantations.

They spent 5 months collecting more than 1,000 breadfruit but less than a month after leaving, a mutiny occurred near Tofua Island in Tonga. His sailors had 'fallen in love' with the Tahitian girls.The sailors, led by Fletcher Christian, mutinied and returned to Tahiti to collect their lady friends and setup a community on Pitcairn Island. They still live there to-day.

Captain Bligh and his supporters were put
into a long boat and set adrift while. Captain
William sailed 6,700 km to Timor arriving
there in late 1789 (Wikipedia).

It is said he was given another boat, the
Providence, to complete the breadfruit
collection and transportation project.
Unfortunately, after all that work the slaves
refused to eat the breadfruit (Wikipedia).

Breadfruit Distribution...

Breadfruit is cultivated on most Pacific islands
and has a pantropical distribution. In the late 1700s,
several seedless varieties were introduced to
Jamaica and St. Vincent from Tahiti and a Tongan
variety was introduced to Martinique and Cayenne
via Mauritius. These Polynesian varieties were then
spread through the Caribbean and to Central and
South America, Africa, India, Southeast Asia,
Madagascar, the Maldives, the Seychelles,
Indonesia, Sri Lanka, and northern Australia.
Breadfruit is now also found in south Florida
(U.S.A.). (Ragone 2006)

Breadfruit is recorded by Ragone (2006) to have been collected from most of the Pacific Islands including Tonga and Samoa for planting in sugar plantations in the islands of the Caribbean during the 17th and 18th centuries. It was widely popular because of its similarity to bread after being roasted on an open fire. The plantation owners, mostly British, thought it would be a cheap food for the slaves in the sugar plantations.

Breadfruit (*Artocarpus altilis*) is a species of flowering tree in the mulberry family, Moraceae originating in the South Pacific and that was eventually spread to the rest of Oceania. Ancestors of the Polynesians found the trees growing in the northwest New Guinea area around 3,500 years ago. They gave up the rice cultivation they had brought with them from Taiwan, and raised breadfruit wherever they went in the Pacific (except Easter Island and New Zealand, which are too cold). Their ancient eastern Indonesian cousins spread the plant west and north through insular and coastal Southeast Asia. wikipedia

## Diane Ragone, Director of the Breadfruit Institute

Dr. Ragone was appointed Director of the Breadfruit Institute in 2003 and has worked at the National Tropical Botanical Garden since 1989. She is an authority on the conservation and use of breadfruit, conducting horticultural and ethnobotanical studies on this important Pacific staple crop for 30 years. Her extensive fieldwork on over 50 islands in Micronesia, Polynesia, and Melanesia enabled the NTBG to establish the world's largest collection of breadfruit at its Kahanu Garden in Hāna, Maui. The Breadfruit Institute was created to promote the conservation and use of breadfruit for food and reforestation. Current research involves collaborative projects to develop *in vitro* methods to conserve and distribute breadfruit cultivars; nutritional and salinity studies; molecular and morphological studies to understand taxonomic relationships, origin, and distribution of breadfruit in the Pacific; and ethnobotanical studies on traditional uses of breadfruit in Polynesia and Micronesia.

Dr. Ragone is an Adjunct Professor at the University of Hawai`i in the Department of Tropical Plant and Soil Sciences. She is the author of more than 80 publications on breadfruit, ethnobotany, horticulture, and native plant conservation. Dr. Ragone holds a Ph.D. and M.Sc. in Horticulture from the University of Hawai`i and a B.Sc. from Virginia Tech.

From 'Staff of National Tropical Garden'.

Figure 28. A dead breadfruit tree, village of Kolovai, Western District, Tongatapu. This area is only 1-2 metres above sea level. Several breadfruit trees were observed to be showing decline in Kolovai. Disease symptoms of fruit rot, 'crown' defoliation and dieback can also be observed. Many young bread fruit trees have lost most of their leaves on the top half. All branches with no leaves are also dead and dry.

During this survey many breadfruit trees were observed to be dying or dead on the Western District of Tongatapu (see Figures 28-36). In the case of the tree in Figure 29 there may a 'salinity' problem as well. The tree is close to the beach area and 'tidal surges' which are associated with climate change and global warming may part of the problem as it increases salinity in the underground water. Pone (2014) has reported that increasing salinity reduces plant growth in the case on *Xanthosoma sagittifolium in vitro.* At 15% salinity growth stops. Plantlets, in the experiment, die at higher concentrations.

In many cases around the Pacific low lying coastal areas are always affected by the 'surging tide'. Many Television documentaries have shown villages in Kiribati and Tuvalu that are invaded by the high tide. The Guardian Newspaper, London have shown similar problems in the Marshall Islands where the sea has invaded coastal towns.

The increasing salinity in the ground water lens may also be a problem for sensitive plants. My experiments at the University of the South Pacific showed that *Xanthosoma sagittifolium*, for example, shows huge reductions in growth at even 5% salinity, *in vitro* (Pone, 2014). Its growth ceased at 15% salinity.

In History...

Sir Joseph Banks and others saw the value of breadfruit as a highly productive food in 1769, when stationed in Tahiti as part of the *Endeavour* expedition commanded by Captain James Cook. The late-18th-century quest for cheap, high-energy food sources for slaves in British colonies prompted colonial administrators and plantation owners to call for the introduction of this plant to the Caribbean. As President of The Royal Society, Banks provided a cash bounty and gold medal for success in this endeavor, and successfully lobbied his friends in government and the Admiralty for a British Naval expedition. In 1787, William Bligh was appointed Captain of the HMS *Bounty*, and was instructed to proceed to the South Pacific for this task.

Wikipedia

Figure 29. A large 'loutoko' breadfruit tree in decline, probably due to disease and underground water salinity. Salinity in the underground water has been shown to reduce plant growth (Pone, 2014). Most of the leaves at the top are gone. They are already dead and dry. The lower branches still have leaves and young fruits developing. This tree is very close to the 'beach'. Similar problems face Tuvalu, Kiribati and the Marshall Islands where the sea have invaded the land in many places. An obvious sign of Climate Change due to Global Warming. Kolovai village, Western District, Tongatapu.

There has been reports in the literature that *Phytophthora tropicalis* and *P. palmivora* has caused rotting of breadfruits in Brazil, Samoa, Micronesia and other parts of the Pacific.No information is available regarding the 'dieback' of the trees themselves, although suggestions have been made that it could be *'Phytophthora'* causing it.

Recommendation...

Urgent studies should be done in Tongatapu to find out how the current 'dieback' problem can be solved as trees on the Eastern District, and all over the island, are also showing signs of 'dieback'.

Scientists say breadfruit trees are dying in alarming numbers across the atolls of the South Pacific.

Scientists say environmental change and the impact of westernisation have contributed to the problem. Science director at Hawaii's National Tropical Botanical Garden, Diana Ragone, says a few years ago the problem was seen as serious, but it had now gone past that. Ms Ragone says on some atolls, people are seeing total eradication of the breadfruit trees. She says while there is no chance of the breadfruit tree becoming extinct, diversity is being lost and in some places whole varieties have disappeared.

Figure 30. This is an example of a healthy breadfruit tree, 'kea' variety, but some of the young fruits have rotted and turned black. They can be seen near the top right hand corner and middle of this picture. It is common that some breadfruit trees have many 'rotting fruits' but no other dieback symptoms are apparent. This is a tree on the roadside at the suburb of Maui, Nuku'alofa, Tongatapu.

Figure 31. An example of 2 breadfruit trees in the advanced stage of 'decline'. Note the dead branches on   top with 'sparse' leaves on the lower branches. These are young trees judging from the size of them, but the disease has completely taken over. Note the sea in the background. Village of Ha'atafu, Western District, Tongatapu.

This area in only 2-3 metres above sealevel. The long 4-5 month drought may have reduced the water lens so much that seawater infiltrated it. Salinity in the ground water is a problem for crops in many Pacific Islands.

Figure 32. This is an example of a healthy breadfruit tree, variety 'ma'ofala'. No symptoms of dieback can be observed. The tree bears a lot of fruits. Leaves are green with no algae or fungal spots. Suburb of Kolomotu'a, Nuku'alofa, Tongatapu.

Ma'ofala is the smallest breadfruit of all the varieties in Tonga. It is also the first to be harvested every season, probably due to its small size. The kea variety has the biggest fruits and are usually last to be harvested every year.

Figure 33. Another example of a declining breadfruit tree, variety 'loutoko'. Note the sparse leaves and dead branches on the top of the tree. This is a typical symptom of the dieback. Some branches have no leaves and are already dead and dry. Village of Kolovai, Western District, Tongatapu.

Figure 34. Evidence of the breadfruit dieback on some breadfruit at the village of Houma, Western District, Tongatapu. The tree in the foreground is just beginning to dieback from the top. The tree in the middle is in advanced decline with most branches bare of leaves, dead and dry. The trees in the background look healthy with no apparent signs of dieback. They may be different varieties. The trees at the front appear to be of the 'puou' variety while the ones at the back appear to be a different variety, probably aveloloa.

*Phytophthora* has been quoted in many literature sources to be the cause of the breadfruit dieback and fruit rot. It is not clear why there is an 'epidemic' at this time. It may be confounded by the recent 5 month drought and increased salinity in the groundwater lens as the freshwater level drops in low lying areas near the sea. Warming temperatures may also play a role.

---

*Phytophthora* infection of breadfruit.

Results of tests confirm the identity of isolate from leaf blight and fruit rot in Brazil as *P. tropicalis* (Cerqueira, A. O.; Luz, E. D. M. N. and De Souza, J.T (2006)).

Previously *P. palmivora* was identified as the agent of fruit rot on breadfruit in Micronesia, Western Samoa and India (Trujillo, 1970; Erwin & Ribeiro, 1996). As *P. tropicalis* was proposed recently as a new species to regroup the isolates of '*P. palmivora* , it is possible that the pathogen reported by the above authors was *P. tropicalis*.

This hypothesis is supported by the pathogenicity tests. This is the first report of *P. tropicalis* as a pathogen of breadfruit in Brazil.

---

Figure 35. The dark spots on this breadfruit is fruitrot
starting on this 'loutoko' variety fruit with outer skin
removed. This fruit was bought from the Talamahu
Market, Nuku'alofa, Tongatapu. The rot was not obvious
until the skin was removed. This is an excellent example
of how this fungal disease spreads on fruits distributed
around the islands. Spores or hyphae within or on the
rotting breadfruit flesh can cause disease in healthy
breadfruit trees. They are so small, they can be carried
by wind or water on to healthy breadfruit trees.

Evidence of the breadfruit dieback was also seen at the village of Tatakamotonga (Mu'a)*, Eastern District in previous years on the variety ' 'aveloloa'. Locals say this particular variety is very susceptible. Both fruits and tree branches are attacked.

The ' 'aveloloa' breadfruit tree affected was only a few years old. It had started fruiting a few years before. The rot first appeared on the fruits in 2012, then the affected branches died. It spread to other branches then the whole tree died a few months later.

Rot of the fruits of the varieties 'loutoko', 'puou', 'aveloloa' and 'kea' are also common (see Figures 30, 35,36 and 37).

* 'Mu'a' refers to the old capital which is made up of several villages including Tatakamotonga, Lapaha and Talasiu.

Figure 36. Rot on the fruit of this 'puou' variety
attributed to *Phytophthora* (see Figures 35 and 37).
The whole fruit slowly rots, turns black and fall off.
Sometimes the rot continue up the branch and
'dieback' while the fruit remain on the branch (see
Figure 37). This tree was found in the village of
Tatakamotonga, Eastern Districts, Tongatapu.

Figure 37. The beginning of 'dieback' on variety ' 'aveloloa'. Note branch with fruit and no leaves and dead branches. This tree still look vigorous and strong. It may take years for the fungus to kill it. This tree was found in the village of Kolovai, Western Districts, Tongatapu.

# CHAPTER 6. YAM DIEBACK (mahunu).

Yams (*Dioscorea alata*) are important staple foods in Tonga and the Pacific Islands. In Tonga, most yams do not seem to be affected by anthracnose. However, the most popular early yam 'kahokaho' is usually the first to show symptoms in a plantation.

Anthracnose is caused by *Collectotrichum gloeosporioides*. It normally attacks the young 'kahokaho' yams during rain showers when spores germinate. The leaves of the yams appear burned hence the name 'mahunu' which means 'scalded or burned' in the Tongan language.

> Anthracnose, caused by the fungus *Colletotrichum gloeosporioides*, is the most serious disease affecting yam (*Dioscorea alata*) in the tropics (Ripoche et al, 2008).

Large 'kahokaho' growers usually spray their crop with a mixture of Benlate 50% wp and Manzate to protect it against the mahunu. Benlate is a systemic fungicide and manzate is a contact fungicide.

From observation, the early kahokaho usually escapes the anthracnose if it is a dry year, similar to tree crops. If plants are vigorous, there is no need to spray regularly. This is an important 'observation' which growers need to take into account before decisions to use chemical control are made.

Another observation worthy of note is the amount of chemical sprays that need to be done during the wet weather.

After the first 2-3 sprays, the kahokaho plants are growing strong and vigorous, there will be no need to spray every 2 weeks.

Spot sprays can be done once a month or so to 'curb' any disease development if it shows up. This will save money on chemicals and other costs.

Figure 38. 'Mahunu' or anthracnose symptoms on popular yam variety 'kahokaho'. Extensive, expanding necrotic areas are always observed on the leaves. The young plants are usually killed by the fungus. Chemical control is usually applied before symptoms appear. Suburb of Halaleva, Nuku'alofa, Tongatapu.

Figure 39. A healthy looking young kahokaho yam.
Note the necrotic older leaves, the beginnings of the
anthracnose disease problem. Plants can recover
with chemical control of anthracnose at this stage.
Suburb of Halaleva, Nuku'alofa. Tongatapu.

Figure 40. This is a typical plot of 'kahokaho' yams (*Dioscorea alata*) with giant taro (*Alocasia macrorrhiza*), intercrops or mixed cropping. Pele (*Hibiscus manihot*) and 'talo futuna' (*Xanthosoma sagittifolium*) are planted on the side of the plot. This is a 2-3 acre plot, about 5 months old and ready to be harvested in 2-3months. Kahokaho, like all early yams, are long and can be up to 7 feet.

## DIEBACK IN THE YAM BELTS OF NIGERIA

Akem (1999) surveyed and isolated *Colletotrichum* and various fungi from yams in Nigeria. Pathogenicity studies prove *C. gloeosporioides* causes yam dieback and necrosis when inoculated alone into 'water yams' (*Dioscorea alata*).

---

SPC Pest Advisory Leaflet No.12

Yam dieback, or anthracnose, is caused by the fungus *Colletotrichum gloeosporioides*. It is probably present in all the countries of the Pacific region and is often a major problem where yams (*Dioscorea spp.*) are grown intensively. Water yam (*D. alata*) is usually more susceptible to anthracnose than other yams.

Long periods of rain favour epidemics of the disease because the fungal spores (conidia) are spread by rain splash. Young foliage is more susceptible to anthracnose. If periods of high rainfall coincide with the young stage of crop development where yam anthracnose is present on infected plants, the disease can quickly spread throughout the crop.

# CHAPTER 7. AVOCADO DIEBACK

Another disappearing fruit tree from Tongatapu is the avocado (*Persea americana*). Once plentiful like mangoes, it is increasingly rare and hard to find. My survey found less than 50 trees as counted from the survey* vehicle.

---

Are there any diseases that seriously threaten the avocado industry?

Root rot and anthracnose are the two main diseases that threaten the Queensland avocado industry. Root rot is caused by the soil-borne fungus *Phytophthora cinnamomi* Rands and, as its name suggests, affects the roots of avocado trees. Anthracnose is caused the fungus *Colletotrichum gloeosporioides* Penz and Sacc (*Teleomorph: Glomerella cingulata*).

www.daff.qld.gov.au

---

* - The survey was carried out by counting the trees of interest from a moving vehicle. The map of Tongatapu (backpages) shows the routes taken by the survey vehicle.

*Phytophthora cinnamomi* is recorded in the literature to attack the avocado roots, killing the whole tree. It will be interesting to find out what is causing the disappearance of this very important food source from the island of Tongatapu, by checking the roots of avocado plants in decline (see Figure 41).

Anthracnose can also cause a major rot and necrosis problem to the fruits and leaves. Some trees were observed to bear lots of fruit, during the survey, but very little leaves. They may be infected and in decline.

*Phytophthora cinnamomi*, the causal agent of avocado phytophthora root rot, attacks the feeder roots, which can result in death of the avocado tree in California. Although the disease has been studied for more than 60 years, definitive control measures have not been found and losses continue to mount. However, many root rot control strategies have been discovered which can reduce the impact of avocado root rot. Integrated management of avocado root rot is key to controlling the disease.

www.californiaavocadogrowers.com

Figure 41. *P. cinnamomi* is recorded to cause death of avocado trees by attacking their roots. Dieback start from the top continues until the tree collapses. Note the 'regrowth' from the base of the trunk and branches on the left of the tree. Suburb of Longolongo, Nuku'alofa, Tongatapu.

Recommendation...

It is urgent to find out which pathogen is causing the dieback of avocado on Tongatapu and find solutions for it.

Figure 42. An avocado tree bearing fruit. Fruits appear to be 'sparse' compared to other trees. Anthracnose affects flowers and fruit set causing young fruit to fall off. This popular tree is in decline probably due to *P. cinnamomi*. Suburb of Longolongo, Nuku'alofa, Tongatapu.

## Signs of avocado root rot *(P. cinnamomi)*.

  The first signs of the disease are observed in the tree canopy , leaves are small, pale green, often wilted with brown tips, and drop readily. Avocado tree shoots die back from the tips, and eventually the tree is reduced to a bare framework of dying branches. Avocado tree death may take from a few months to several years, depending on soil characteristics, cultural practices and environmental conditions.

www.californiaavocadogrowers.com

## Integrated Management of the Disease

*Phytophthora cinnamomi* can be controlled through good management practices. The MAFFF, Tonga should;
1. Recruit experts to identify and study the disease. A PhD or MSc student would be the best and cheapest way.
2. Implement the recommended control measures.
3. Replant avocado trees in all the villages.

Figure 43. One of the very few healthy and fruiting avocado trees, growing wild at a cattle farm paddock. Village of Kolovai, Tongatapu.

Varieties of Avocado...

I can remember, at least, four varieties I have eaten as a youngster. There may be more.

One variety was so big it resembled a small 'candy red' water melon. Unfortunately, this variety may have disappeared already.

# CHAPTER 8. BANANA BUNCHY TOP VIRUS

Banana export was big business for Tonga in the late 1960s. Export to New Zealand was more than 20,000 tonnes during the years of 1967,68, 69. It was a huge source of foreign currency for Tonga. The introduction of Black Leaf Streak (*Mycospharella fijiensis*) which attack the banana leaves, caused such a huge problem that production dropped from 20,000 tonnes in 1967-69 to as low as 500 tonnes in the 1980s. Part of the problem was the introduction of mono culturing in the 1960s which allowed the BLS to spread quickly through the plantations due to the proximity of plants to each other. Fungal spores are blown or splashed across to healthy leaves where they germinate and penetrate the leaf causing necrosis of the banana leaves. In the worst cases all the leaves 'dry up' due to the necrotic tissue with just 1-3 leaves left at the top. A healthy banana plant can have up to 15 or more green leaves.

Despite a $5 million 'rescue' project by the New Zealand Government in the mid 1980s, the Banana Industry never recovered and the decline of export due to disease, high production and export costs finally caused its ultimate demise in the 1990s.

Banana Bunchy Top Virus was one of the important economic constraints to banana production during the export years and it is still a problem on banana and plantain for subsistence farmers. Infected plants or suckers are left in old 'fallow' plots which become the source of BBTV, spread by migrating winged aphids to infect newly planted banana and plantain.

Banana Bunchy Top Virus

Banana bunchy top virus (BBTV) is one of the most serious diseases of banana. Once established, it is extremely difficult to eradicate or manage. BBTV is widespread in Southeast Asia, the Philippines, Taiwan, most of the South Pacific islands, and parts of India and Africa. BBTV does not occur in Central or South America.

Co-operative Extension Service, College of Tropical Agriculture and Human Resources, University of Hawaii at Manoa. www.issg.org

Once a stand of banana or plantain is infected by BBTV, all of the plants will eventually be infected. It usually spread rapidly by the banana aphid, *Pentalonia nigronervosa,* which is very difficult to contain or eradicate. The aphid life cycle include a 'winged phase' which migrate and 'colonize' healthy banana and plantain plants thus spreading the disease.

Data from research at the MAFF Research Station, Tonga, in the late 1980s, suggest that *P. nigronervosa* develop wings and migrate in search of new hosts. It is 'triggered' by 'overpopulation' on the existing host. The number of aphids on a banana or plantain plant therefore determines when the 'winged phase' develop (Stechmann, pers. comm.).

Recommendation...

MAFFF, Tonga should actively promote control measures of BBTV and replanting of banana and plantain in farmer's fields. BBTV infected mats have to be destroyed with weedside or chopped up.

Figure 44. This is a 'mat' of banana bunchy top virus infected plantain or 'hopa' suckers. Note the stunted growth and 'bunched' leaves. The 'parent' plant has been removed. BBTV has become 'systemic' within the plant. All new growth will be infected. This is the reason why growers are advised to remove 'all BBTV infected' plants by digging out the whole mat and destroy them or the suckers will become 'wild sources' of infected winged aphids which will infect new healthy plantations of bananas or plantain. These suckers will act as 'wild' sources of BBTV infecting newly planted plantain and bananas in this growers plantation. Village of Kolovai, Western District, Tongatapu.

Figure 45. The suckers on this mat all show
symptoms of BBTV.   Although the parent plants
are healthy, all subsequent growth will be stunted
with bunched leaves due to BBTV infection. Village
of Kolovai, Western District, Tongatapu.

Aphid control...

Infected mats should be sprayed with an insecticide,
to kill the aphids, before the whole mat is destroyed
with a suitable weedicide or chopped up.

Figure 46. An older plant beginning to show symptoms of erect, bunched leaves due to BBTV infection. Village of Kolovai, Western District, Tongatapu.

Figure 47. A single sucker emerged from an
infected plant. Once a single mat is infected, the
rest of the plantation are infected by migrating
winged aphids. Village of Kolovai, Western District,
Tongatapu.

# CHAPTER 9. SOOTY MOULD

This is a very common problem that has come up in my discussions with the agriculture staff in the Eastern District of Tongatapu. Many plant species has a black mould growing on their leaves hence the 'sooty mould' name. It is so bad in some cases the whole plant looks black.

It is also very common. This aggressive growth of the mould is only a recent phenomenon. It has never been observed to be this common and widespread in the past.

Sooty mould is caused by many fungi in association with an exudate produced by insects living on the affected plant leaves. The mould grows on the exudate.

Many species of fungi are recorded in the literature to cause sooty mould include *Cladosporium, Aureobasidium, Antennariella, Limacinula, Scorios, Capnodium*. The fungi grow on the sugary exudate of the insects causing the

black 'sooty' appearance and can cause
problems with plant photosynthesis. If there
is no exudate there will be no 'soot'.
In Tonga sooty mould has been observed on
mangoes (*Mangifera indica*), koka (*Bishofia
javanica*), citrus (*Citrus sp*), Siale (*Gardenia
jasminoides*), heilala (*Garcinia sessilis*)
and possibly other species.

Figure 48. A mango tree totally covered with sooty
mould at the village of Kolovai, Western District,
Tongatapu. Note the new green shoots on the right
of the tree.

There is no treatment recommended for this problem. Old leaves will fall off and new 'clean' ones will emerge (see Figure 48 and 50). There is a new scale insect associated with plants that have the mould. It should be interesting to find out if the scale is producing the exudate for the mould to grow on.

SOME PLANTS AFFECTED BY SOOTY MOULD.

Figure 49. Sooty mould on mango. Note the scale insects on the mango leaf midrib. Suburb of Kapeta, Nuku'alofa, Tongatapu.

Figure 50. Sooty mould covering a 'moli vaikeli' plant. Note the new green leaves are not affected by the mould. Suburb of Longolongo, Nuku'alofa.
Figure 51. A young rough lemon affected by sooty mould. Note the young 'clean' leaves. Suburb of Hala'ovave, Nuku'alofa.

Figure 52. A Heilala plant affected by sooty mould. Note scale insects on leaf mid-rib. This large scale is probably a new pest. The author has never seen it before in Tonga. Suburb of Pahu, Nuku'alofa.

Sooty Mould Control

Sooty mould grows on an exudate from an insect. There is no urgent reason to control it. New leaves usually develop without the mould as old affected leaves fall from the tree.

Figure 53 and 54. A lime tree, probably a natural
hybrid, affected by sooty mould. Suburb of
Tufuenga, Nuku'alofa.

# CHAPTER 10. TONGA'S FOOD SECURITY

This article makes interesting reading in light of what is happening in Tonga.

World Rainforest Movement, Oceania and Pacific.
Deforestation and Forest Degradation in the Kingdom of Tonga
Denis Wolff, Executive Director,
Tonga Community Development Trust

The Kingdom of Tonga has experienced significant deforestation and forest degradation. Two primary causes for this are identified. The first is population change; most importantly the rapid and substantial increase in population during the past century, with its associated impacts of increased/intensified land-use, and decreasing availability of land. The second is economic change; most importantly the monetization of Tonga's economy, with its associated impacts of increased need for disposable income, commercialisation of agriculture to meet this need, and consequent increase and intensification of land-use. The overall impact has been a decline in Tonga's tree and forest resources.

A number of contributing factors and/or obstacles are identified. These include land tenure, agricultural and forestry policy, changes in agricultural methods/practices, changes in dwelling patterns & associated urbanisation, changes in human attitudes, rapid pace of change, and contradictions between relevant sectors and applicable policy.

www.wrm.org.uy

Some issues of food security in Tonga are discussed in this chapter. Probably because it has never been raised. This short survey does raise this issue due to these facts;

1.    Important citrus fruit trees, which are sources of vitamin c for the villages, all over Tonga have been lost due to viral disease.

Two very important citrus species 'moli kai' (*Citrus sinensis*) and 'moli peli' (*Citrus reticulata*) are in this category. Although they were popular in the 1970s, with large amounts being sold at Talamahu Market, sports tournaments and elsewhere, they have completely disappeared due to citrus tristeza virus (CTV).

2.    Other important fruit trees, like the avocado, lime and rough lemon, are in decline. They are not as common as they used to be.

3.    There are reports of pawpaw disease destroying 'plantations' of pawpaw. This is a major concern,

4.    Breadfruit varieties all over Tongatapu exhibit symptoms of dieback due to a disease suspected to be *Phytophthora*

*palmivora* or *P. tropicalis.*

5.    Long droughts are more frequent, suspected to be a 'symptom' of Global Warming.

6.    Coastal areas, like in the village of Kolovai, Western District are affected by 'sea surges' or high tides which affect fruit trees like breadfruit (see Figure 29). Also a symptom of global warming and climate change.

7.    <u>Destructive diseases, like cassava mosaic virus may be introduced, accidentally, in the near future. This will cause a huge problem to the local food supply. Cassava is one of the major local staples.</u>

8.  Reduction in production of some food species due to plant diseases.

Many food species fit in this category. They still produce but are greatly affected either in reduced quality or quantity. Mangoes (anthracnose), avocado (anthracnose), breadfruit (*Phytophthora*) and yams (anthracnose) are some examples.

9. Loss of some food species due to overexploitation.

This point is worth mentioning, although it is not a plant protection concern. These seafood species were once plentiful. Now they are scarce or have disappeared.

A. Sipesipa (*Leiognathus equulus* or ponyfish). This species of fish, once plentiful, have disappeared from Tongatapu reefs.

B. To'o – (*Gafranium tumidum*). Cockles are rarely seen in the fish market as they have also disappeared or greatly reduced in numbers from the lagoon and surrounding islands.

C. Kaloa'a ( *Andara spp* or Ark clams). Similar to cockles, this species is also disappearing from the lagoon and surrounding islands. Lui et al (1994) suggest that decreasing amounts of this shellfish already occurred prior to its 'modern exploitation'.

D. Lomu (*Stichopus variegatus*) – Sea cucumber and sea slugs are becoming rare, probably due to overfishing.

E. 'Ufu (*Leptoscarus vaigiensis*) – A popular reef fish. Once found in large numbers on the Nuku'alofa waterfront, have become rare.

F. Vasua (*Tridacna spp*)– Giant clams are also declining in numbers. Although they can still be seen at the local fish market, the high prices suggest they are harder to find.

---

The Disappearing Jellyfish...

Once plentiful in the lagoon the jellyfish species known as 'ma'anu', because it floats on the surface', is rarely seen now. Some people say it is seasonal but I remember when we used to fish in the lagoon as kids, our net was full of it all the time. Only the 'toka' species, because it sinks to the bottom, is still commonly sold on the roadside or at Talamahu Market in Nuku'alofa. .

---

*Tridacna derasa* (smooth giant clam - vasuva (tokanoa molemole), *T. maxima* (elongated or rugose giant clam - kukukuku).

'Today giant clams come under increasing pressure because of the increased levels of exploitation to supply several types of demands. They are still harvested on a subsistence basis but have become an important component in the artisanal fishery, where they are targeted for the domestic market, especially in Nuku'alofa'.

Fisheries Resources Profiles, Kingdom of Tonga. Prepared by Lui A. J. Bell, 'Ulunga Fa'anunu and Taniela Koloa. FFA Report 94/05.

These are just naming a few, as examples, to emphasize the importance of securing the food supply by better management of local resources. Better management can help revive the population of these important food species, but it may take a long time.

Some species cannot survive below a certain number of reproductive individuals.

Deforestation and Forest Degradation in the Kingdom of Tonga by Denis Wolff. Executive Director, Tonga Community Development Trust. www.wrm.org.uy

Trees and forests have consistently been relegated to a very minor position within agricultural policy. Although the traditional Tongan farming system was essentially an agroforestry system, scant mention of trees or forests is made in the agricultural policy sections of Tonga's various development plans. While acknowledging that "there is always potential for development of tree crops", the actual practice has been "to utilise land unsuitable for agriculture for forestry development".With the exception of passing references to lime trees, paper mulberry, and pandanus, trees and tree crops are mentioned only in relation to coconut and banana, and within the context of export of copra, coconut oil, and banana. It is interesting to note that none of these three currently play any significant role within the framework of commercial agriculture. While the world market value of coconut products has increased somewhat in the recent past, it is still far below potential income earned from other cash crops. As for banana, this is a useful example of commercial promotion of an agricultural crop that ultimately proved to be of questionable sustainability. Bananas proved to be sustainable only so long as subsidies were provided (through various banana export schemes) and a protected market existed. Once the subsidies ended, and when the New Zealand market was deregulated, exporting of bananas from Tonga came to a virtual standstill.

## CAN TONGA AFFORD TO IMPORT ALL ITS FRUITS?

The balance of trade as it is, does not give much hope. The export figures is around $30 million while import is $300 million (Tonga Stats. 2012). The people who can influence this change for the better or worse, like the Minister of Agriculture, can put certain policies in place to protect Tonga's fruit trees from pest and diseases.

## Food Security...

Food security must be discussed at the highest level and recommendations from the experts implemented to avoid food supply problems in the future.

There must an active tree crop replanting and protection programme added to the MAFFF future plans.

# CHAPTER 11. PLANT PROTECTION MANANGEMENT.

It is obvious from the discussion with local staff and the evidence and pictures collected for this book that there are some serious disease problems on tree crops on Tongatapu. These trees include citrus, avocado, breadfruit and pawpaw. Other tree crops like tava (*Pometia pinatta*), fekika ( *Syzygium malaccense)* and ifi (*Inocarpus edulis*) should also be checked.

Kava dieback is still very common. It is obvious from the plants found that CMV is present. Kava dieback was devastating before (Davis, Brown and Pone, 1996). It is still a serious problem for kava growers. Control measures recommended by Davis et al (1996), Davis (1999) and Davis (2005) need to be reinforced. I have included some recommendations for CMV control in kava plantations in this book (see Chapter 4). The few vanilla plantations inspected did not show any VNPV symptoms. It is evidence that control measures designed in the past are still working (Pone, 1988; Pone and Pearson, 1988).

Some kind of plant protection management plan should be put in place for the long term to protect the tree crops of Tonga from the disease or diseases affecting them.

<u>Most fruit trees in Tonga regenerate naturally with a small percentage that are planted by growers. Trees are usually left to nature to nurture them. However, it is increasingly important for the tree owners to manage the trees, especially plant protection measures.</u>

MANAGEMENT STRATEGIES.

1. A comprehensive survey need to be done urgently to determine the exact state of the tree crops on Tongatapu, other main island groups, species and numbers of individuals in existence. This information will be very useful in making decisions about plant

protection management.

2. Diseases affecting trees on Tongatapu and other main island groups in Tonga need to be identified and confirmed.

3. Studies of the diseases affecting trees in Tonga should be carried out at MSc and PhD level to collect enough information for future disease control/plant protection decisions and train local experts.

4. Once all the information is available, long-term disease control strategies can be designed and the growers advised.

5. The Plant Protection series of books, available from amazon.com, including this book contain enough information to help guide the management planning for future agricultural prosperity through better plant health.

6. The MAFFF should implement the Plant Protection Management Programme as a matter of priority.

# CHAPTER 12. CLIMATE CHANGE

Some programmes in the Pacific Islands try to address the effect of climate change on agriculture.

## Pacific Climate Change Science

Small island developing countries are among the most vulnerable to climate change. During this century, these countries will face increasing threats to sustainable development from the impacts of climate change. Sectors which are likely to be most affected include human health, infrastructure, coastal resources, disaster management, fresh water availability, agriculture, fisheries, forestry, marine ecosystems and tourism. The Pacific Climate Change Science Program (PCCSP) was funded by AusAID, managed by the Department of the Environment (DOE) and delivered by a partnership between the Australian Bureau of Meteorology and Commonwealth Scientific and Industrial Research Organisation (CSIRO) during the period 2009- 2011 to help the Pacific Islands. http://www.pacificclimatechangescience.org

The Secretariat of the Pacific Regional

Regional Environment Programme (SPREP) is also very much involved in this area.

It would be interesting to try and link the problems facing food species in the Pacific Islands with global warming. Pests and diseases may become more problematic as we are seeing now. For example, the rhinoceros beetle has reappeared after disappearing for years due to virus biocontrol. Damaged trees are now very common (See Figure 58).

Species of important food plants like some citrus species are disappearing from some islands, varieties of breadfruit are disappearing from others . It may take some time of concentrated scientific research to make this link.

Global warming is a catalyst in the disappearance of food species from the islands because of more 'aggressive' pests and disease. There is evidence in the island of Tongatapu, Kingdom of Tonga that decline in the number of some fruit trees is happening but the locals do not even notice!

Such is the effect of global warming that large scale changes of plant and animal species in the islands may occur over time. In many cases it may affect the health of the population due to insufficient supply of the right foods . It is already happening.

Important sources of vitamins like citrus are disappearing from Tongatapu and the outer islands of Tonga perhaps it was a slight increase in temperature that caused greater aphid activity which resulted in a massive explosion and death of tristeza infected trees?

Such large scale changes in the species composition of island countries will probably become more and more common now. Global warming may be the cause of greater disease and pest activity that are killing off citrus, breadfruit, avocado and pawpaw. Usually epidemics are caused by changes in the environment favoring the disease or insect development.

In the case of taro in American Samoa and Samoa warming temperatures and wet weather allowed the fungus *Phytophthora colocasiae* to wipe out all the taro on both islands in less than a month!

Both Samoas are still struggling to revive their taro industries after it was destroyed by taro leaf blight (TLB) in 1993. Breeding programmes for resistance against TLB have been successful but due to various reasons, growers cannot export taro yet.

I visited both Samoas in 1993 to view the damage to their taro. It was unbelievable. I have never seen whole plantations of taro with their leaves all burned as if scalded by hot water! Hundreds of plants in some cases with large necrotic spots on the leaf lamina. Some experts believe it is a 'new' more virulent strain. This 'change' in fungal genetic makeup can only happen with selection pressure on it caused by environmental factors.

The Pacific Adaptation to Climate Change (PACC) Programme is the first major climate change adaptation initiative in the Pacific region.  Since it began in 2009 the Programme has been laying the groundwork for more resilient Pacific communities that are better able to cope with climate variability today and climate change tomorrow. The Programme approaches this from two directions: it is working to enhance adaptive capacity on the ground, and it is driving the mainstreaming of climate risks into national development planning and activities.

Working in 14 Pacific island countries, the Programme is *demonstrating best-practice adaptation* in three key climate-sensitive areas: coastal zone management, food security and food production, and water resources management. Each country is hosting a pilot project in one of these theme areas to demonstrate how climate change adaptation can work on the ground.

Climate change threatens the achievement of all development goals. One solution is to mainstream climate change into the development process, that is, integrate climate risks into development planning processes and decision making.

www.sprep.org

Apart from the effect of climate change on coastal regions and so on, the changes in species composition due to the effect of more virulent diseases will become more pronounced in future.

Figure 55. Warming temperatures will affect the growth of kava and also promote greater aphid activity which will spread    Cucumber Mosaic Virus with devastating effect. Whole plantations have succumbed to the virus in the past and will probably become more frequent in future if CMV control is not strictly enforced.

Figure 56. With warming temperatures, increased aphid activity will spread pawpaw diseases with devastating results. New diseases will become more common and more devastating as the warming environment favors their development.

Did you know?
Both green papaya fruit and the tree's latex are rich in papain, a protease used for tenderizing meat and other proteins. Its ability to break down tough meat fibers was used for thousands of years by indigenous Americans. It is now included as a component in powdered meat tenderizers. wikipedia.

Figure 57. This breadfruit tree is showing signs of decline/dieback. Notice the sparse leaves and lack of fruits. Such dieback symptoms have been attributed to *Phytophthora palmivora.* The roots of this 'loutoko' breadfruit variety at Kolovai, Tongatapu should be dug up and examined a bit closer. It could also be a result of warming temperatures as many other breadfruit in this area are showing signs of decline/dieback which is a sign of greater fungus activity due to the warming temperatures.

Figure 58. Damage on coconut leaves due to rhinoceros beetle. This pest was under control through biocontrol 20 years ago, but has now reappeared. Sometimes when pests are in very low numbers biocontrol agents population crash through lack of hosts, then the pest population recovers. Or is it the influence of warming temperatures? Eastern District, Tongatapu.

# NOTES ON SURVEY.

## A. 'QUICK ASSESSMENT' OF TREE CROPS ON TONGATAPU.

In this exercise the number of fruit trees beside the main road from (see Map of Tongatapu, backpages.), from the village of Niutoua on the Eastern District to the village of Ha'atafu in the Western District of Tongatapu, were counted from a moving vehicle. The objective is to get an 'impression' of the abundance of the fruit trees, without worrying about the accuracy of the numbers. The premise is that, if you ride in a vehicle at 40km per hour and watch the trees by the road as you go past, you will rapidly get an impression of the relative numbers of the trees. For example, coconuts are more abundant than mangoes. Mangoes are more abundant than avocado. Tava trees are more abundant than vi trees and so on. These numbers are useful in making a 'quick' assessment of the 'collective' health of each species. When a diseased tree is spotted, the truck stops and photos are taken for the records. Thus this 'quick survey' gives a very good assessment of the overall health of each species of trees. When the relative numbers are compared with disease records, it becomes clear what is happening.

The photographic records of the species and severity of each disease (see photos), supports the 'impression' from the survey. Trees that are not abundant now but were before, like avocado, are being attacked by diseases.

Information from the survey was compared with the scale based on 'abundance' (see Table 2) as a 'rough' guide to the state of the fruit trees on Tongatapu. The results are in Table 3 with disease records in Table 4.

The major assumption is that;
If fruit trees are 'abundant' then there is no problem affecting them.

The 'abundance' of the trees in this survey is compared with other information, like reports of disease in the literature, to determine whether it is 'thriving' or 'declining'. If a known disease is reported to affect that species and it is in 'decline' then we can be sure there is a problem.

The 'abundance' of these fruit trees is cause for concern when plant protection information is included as most trees in 'decline' is affected by an insect pest or disease.

Normally there is an abundance of fruits sold on the roadside markets when in season. In this sense 'availability' or sale on the roadside can be used as an indicator of 'abundance'. Some fruits are also sold at the Talamahu Market in Nuku'alofa. If the fruit is not abundant compared to a few years ago then it is a good reason to find out why.

This comparison is based only on the author's experience and what he has seen. For example, how many sweet oranges and mandarins were sold in 1975 compared to 2015. In 1975, there were an 'abundance' of sweet orange and mandarin sold in

Nuku'alofa. Now, none can be seen or found in Nuku'alofa, the villages or during the 'quick survey' . No sweet orange and mandarin trees were found or seen during this survey, either.

| Table 2. A scale to measure tree crop decline from a moving vehicle on the road. |
| --- |
| 0       - no trees found or seen<br><5      - trees rarely found or seen<br><50    - very few trees found or seen<br>>100   - a small number of trees found or seen<br>>1,000 - trees found or seen are 'abundant' and are too numerous to count. |

This is a 'rough and quick' way to check the tree population and any possible disease problems. Once this information is gathered, a comprehensive survey is carried out depending on whether a series disease situation is detected. Laboratory tests, overseas 'professional organisation' confirmation of disease species and MSc, PhD programmes follows. This step may take years to complete. It will also ensure there are 'experts' on the problem with information on what to do.

| Table 3. Fruit trees and their 'abundance' based on this survey method. | |
|---|---|
| 1. Mango | >1,000 |
| 2. Tava | >1,000 |
| 3. Breadfruit | >1,000 |
| 4. Moli kai | 0 |
| 5. Moli peli | 0 |
| 6. Moli vaikeli | >100 |
| 7. Moli Tonga, moli uku | <5 |
| 8. Lemani,kola | <50 |
| 9. Avocado | >100 |
| 10. Coconuts | >1,000 |
| 11. Vi | >100 |
| 12. Fekika | <5 |
| 13. Kuava | >1,000 |
| 14. Pawpaw | >1,000 |
| 15. Ifi | >100 |
| 16. Banana | >1,000 |
| 17. Hopa | >100 |

The numbers tell a very clear story (Table 3). Some species of fruit trees are still abundant, others have disappeared or are 'rarer'. Some are declining and will probably disappear in the next few years, if nothing is done about it.

If you add the pests and diseases the picture becomes very grim indeed as fruits like avocado may disappear in the next decade.

WHAT CAN BE DONE?

It is time for Tongan farmers to take responsibility for management of tree crops to avoid the problem of losing some fruit species completely. They cannot be left to regenerate naturally, there must be an active replanting and disease management programme.

Information can already be collected in the first year for designing the Plant Protection Management Programme.

Table 4. Fruit trees with important disease and pest problems.

1. Mango - anthracnose
2. Tava - anthracnose
3. Breadfruit - *P. palmivora*?, *P. tropicalis*?.
4. Citrus- tristeza, psorosis/fruit flies*.
5. Coconuts - rhinoceros beetle+
6. Vi - no major disease/fruit flies*
7. Fekika - no major disease/fruit flies*
8. Kuava - no major disease/fruit flies*
9. Ifi - no major disease
10. Banana - banana bunchy top virus
11. Pawpaw - 'yellowing disease'
12. Hopa (plantain) - banana bunchy top virus
13. Avocado - *P. cinnamomi*?, anthracnose.

* - although there are no major disease on some fruits, fruit flies are a major problem on the ripening fruits. Fruits infested by fruit flies are often inedible.

+ - rhinoceros beetle damage disappeared when biocontrol was used by the Tonga-German Plant Protection Project from 1982-1992 (approx.). Now , 22 years later, the damage has increased judging from the number of trees attacked.

If tree crops can be better managed the decline can be reversed and new virus resistant 'moli kai' (sweet orange) and 'moli peli' (mandarin/tangelo) can be introduced.

## THE NUKU'ALOFA SURVEY.

The survey of Nuku'alofa was done on foot. The author feels that it will be better to take time and look at each plant. More than a 1,000 pictures were taken of plants and flowers of Nuku'alofa. The idea was to check what plants Nuku'alofa residents have in their properties, and keep a record of them.

## THE KAVA AND VANILLA SURVEY.

All plantations and kava plants found on Tongatapu were assessed in terms of disease symptoms. Pictures taken show the severity of the CMV dieback disease on kava.

The vanilla survey was an inspection of the plants in 5 plantations for VNPV symptoms and to take pictures of them.

Loss of Diversity....

There is plenty of evidence to support this statement. Diana Ragone of the Breadfruit Institute, Hawaii had mentioned that in the case of breadfruit many varieties have been lost from islands around the Pacific.

Certainly in the case of Tonga *Citrus sinensis* (sweet orange) and *Citrus reticulata* (mandarin) have definitely been lost. Many other fruit trees are in decline.

Fish species such as sipesipa has disappeared from the reefs around Tongatapu. Other shellfish are in decline or rarely found.

A huge effort is needed to turn the decline and loss of species around.

# NOTES ON THE AUTHOR.

Semisi Pone was a Plant Pathologist/Agriculture Officer for the Ministry of Agriculture, Fisheries and Forests in Tonga after graduating with a BSc (1985) and MSc (Hons) (1989) from the University of Auckland.

He was promoted to the post of Senior Plant Virologist, MAFF Research Division, Tonga in 1991 for his work on vanilla, kava and squash virus diseases.

He joined the Institute for Research, Extension and Training in Agriculture (IRETA) at the University of the South Pacific, Alafua Campus, Apia, Samoa as a Fellow in Tissue Culture in March 1992. During his time at USP he was involved with storage and dissemination of tissue cultured crops for research around the Pacific Islands.

In May 1993, he was appointed the Plant Protection Advisor for the South Pacific Commission based in Suva, Fiji. He was also the Co-ordinator for the SPC Plant Protection Service which had more than $20 million in projects and more than 20 staff around the Pacific Islands.

The establishment of the Pacific Plant Protection Organisation was one of Mr Pone's most important tasks. It has been under discussion for 8 years previously by the 27 member countries of SPC. Assistance from the Food and Agriculture Organisation of the United Nations Legal Advisor, Mr Richard Stein, and the Ministry of Foreign Affairs staff, New Zealand enabled the drafting of a resolution for the establishment of the PPPO. The 34th South Pacific Conference in Noumea, New Caledonia approved and passed the resolution establishing the PPPO in October, 1994.

Mr Pone also arranged for a Memorundum of Understanding (MOU) to be signed between the newly formed PPPO and the

Asia Pacific Plant Protection Commission (APPPC) to work closely together in the Asia-Pacific region. The Director of Programmes Mr Poloma Komiti, representing the SPC Secretary General, and Professor Shen, APPPC Executive Secretary, representing the APPC signed on behalf of the Director General of FAO and the APPPC.

Other work that Mr Pone was involved in include;

1. Manager of the SPC/EU Pacific Plant Protection Project with a budget of $5 million Fijian dollars.

2. Member of the Committee of Experts on Phytosanitary Measures (CEPM) and Regional Plant Protection Organisation (RPPO) Technical Consultation which meet at FAO HQ in Rome every other year.

He moved to New Zealand in 1996 and was involved with many businesses for 15 years.

He decided to become a writer in 2011 and
is now a full time writer of children's stories,
poetry, humor, novels and science books.

# REFERENCES.

1. Bell, L. A. J. , Fa'anunu, 'U. and Koloa, T. (1994). Fisheries Resources Profiles, Kingdom of Tonga. FFA Report 94/05.

2. Cerqueira, A. O.; Luz, E. D. M. N. and De Souza, J.T (2006). First record of *Phytophthora tropicalis* causing leaf blight and fruit rot of breadfruit in Brazil. Plant Pathology Volume 55, Issue 2, page 296.

3. Davis R. I., Lomavatu-Fong, M. F. ; McMichaef, L. A; Ruabete, T. K.;   Kumar, S. and Turaganivalu, U. *Cucumber mosaic virus* infection of kava (*Piper methysticum*) and implications for cultural control of kava dieback disease. Australasian Plant Pathology. Volume 34, Issue 3. Pp 377-384.

4. Davis, R.I. (1999). Pest Advisory Leaflet No.25, Plant Protection Service. Secretariat of the Pacific Community, PM Bag, Suva, Fiji.

5. Davis, R.I. (SPC), Lomavatu-Fong,M; Ruabete, T and Turaganivalu, U. (Ministry of Agriculture, Sugar and Land Settlement; Fiji), (2005). Integrated Management of Kava Dieback Disease (2005). Pest Advisory Leaflet No.47, Plant Protection Service, Secretariat of the Pacific Community, PM Bag, Suva, Fiji.

6. Davis, R. I.; Brown, J. F.; Pone, S. P. (1996). Causal relationship between cucumber mosaic cucumovirus and kava dieback in the South Pacific. Plant Disease Vol. 80 No. 2 pp. 194-198.

7. Erwin DC, Ribeiro OK, 1996. *Phytophthora Diseases Worldwide*. St Paul, MN, USA: APS Press.

Holo, Tevita F. (1985). Anthracnose of mangoes and treecrops. Principal Plant Pathologist, Vaini MAFF Research Division, Tongatapu, Kingdom of Tonga. Personal Communications.

8. Kanmwischer ME, Mitchell DJ, 1978. The influence of a fungicide on the epidemiology of black shank of tobacco. *Phytopathology* 68,1760–5.

9. Manu, V. (2015). Pawpaw disease. MAFFF Deputy Director, Kingdom of Tonga. Personal communication.

10. Mossop D.W. and Fry P.R. (1984). Records of viruses pathogenic on plants in Cook Islands, Fiji, Kiribati, Niue, Tonga and Western Samoa. South Pacific Bureau for Economic Cooperation/United Nations Development Programme/Food and Agriculture Organization of the United Nations

11. Pone, S.P. (1988). An investigation of 3 virus diseases of *Vanilla fragrans* (Salisb.) Ames in the Kingdom of Tonga. MSc Thesis. University of Auckland.

12. Pone, S.P. And Pearson, M.N. (1988). Koe mahaki vailasi 'o e vanila. Advisory brochure. MAFFF, Tonga.

13. Pone, S.P. (2013). Plant Protection in the Pacific. Rainbow Enterprises, Auckland, New Zealand. 114 pp.

14. Pone, S.P. (2014). Plant Protection in the Pacific 2. Rainbow Enterprises, Auckland, New Zealand . 78 pp.

15. Pone, S. P. (2014). Plant Protection in the Pacific 3. Rainbow Enterprises, Auckland, New Zealand. 126 pp.

16. Ragone, Diane (April 2006). "*Artocarpus altilis* (breadfruit)" (PDF). The Traditional Tree Initiative. Archived from the original on 28 December 2013 (Encyclopedia of Life).

17. Ripoche, A.; Jacqua, G.; Busierre, F.; Guyader, S and Sierra, J. (2008). Survival of *Colletotrichum gloeosporioides* on yam residue decomposing in soil. *Applied Soil Ecology*. Vol. 38, Issue 3. Page 270-278.

18. Samuela (2015). Kava harvests. MAFFF staff, Eastern Districts, Tongatapu, Kingdom of Tonga. Personal communication.

19. Stechmann, D. (1986). Banana aphid. Entomologist and Manager of the Tonga-German Plant Protection Project, Vaini Research Station, MAFFF, Tongatapu, Kingdom of Tonga. Personal communication.

20. Trujillo E.E., 1970. A *Phytophthora* fruit rot of breadfruit. *Phytopathology* 60, 1542.

21. Wikipedia. The free online encyclopedia.

# MAP OF TONGATAPU

Map show roads taken and villages visited during the surveys.

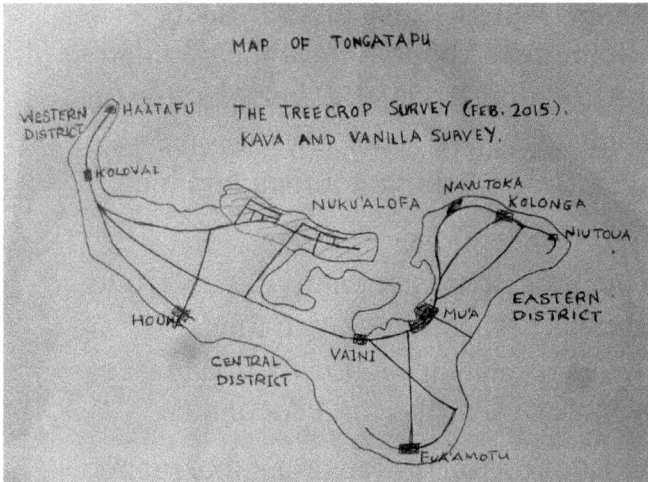

## TONGATAPU...

Tongatapu is the main island of the more than 170 islands of the Kingdom of Tonga. It is 99 square miles or 159 square kilometres. About 70% of the population of the kingdom or 71, 260 persons live on Tongatapu. The main activities are agriculture, fishing, tourism and local industries.

The main crops are yams, cassava, talo futuna (*Xanthosoma*), talo tonga (*Colocasia*), kape (*Alocasia*), bananas, plantain, breadfruit and coconuts.

Most of the other tree crops like mangoes, tava (*Pometia*), avocado, vi (*Spondias*) and so on regenerate naturally but are very important part of the Tongan people's diet.

It is important to replant the 'lost trees' and 'declining tree species' and also manage the diseases that affect them.

# PLANT NAMES.

| Tongan | English | Scientific |
|--------|---------|------------|
| 'Avoka | Avocado | *Persea americana* |
| Fekika | Mountain apple | *Syzygium malaccense* |
| Heilala | | *Garcinia sessilis* |
| Ifi | Polynesian chestnut | *Inocarpus edulis* |
| Kape | Giant taro | *Alocasia macrorrhiza* |
| Kava | | *Piper methysticum* |
| Laimi (Kola) | Lime | *Citrus latifolia* |
| Lemani | Rough lemon | *Citrus limon* |
| Lesi | Pawpaw, papaya | *Carica papaya* |

| | | |
|---|---|---|
| Mei | Breadfruit | *Artocarpus altilis* |
| Moli kai | Sweet orange | *Citrus sinensis* |
| Moli peli | mandarin | *Citrus reticulata* |
| Moli tonga | Pomelo | *Citrus grandis* |
| Moli uku | Sour orange | *Citrus aurantium* |
| Niu | Coconut | *Cocos nucifera* |
| Pele | Polynesian spinach | *Hibiscus manihot* |
| Siaine | Banana | *Musa sapientum** |

| Talo futuna | Taro tarua | *Xanthosoma sagittifolium* |
| --- | --- | --- |
| Talo tonga | Taro | *Colocasia esculenta* |
| Vi | Vi-apple | *Spondias dulcis* |
| 'Ufi | Water yam | *Dioscorea alata* |
| Vanila | Vanilla | *Vanilla fragrans* |

*- FAO in its ECOCROP classifies hopa and siaine as *Musa sapientum*. They may be different cultivars of the same species!

# ABBREVIATIONS.

1. BBTV - Banana Bunchy Top Virus

2. BSc - Bachelor of   Science

3. CEPM - Committee of Experts on Phytosanitary Measures, FAO.

4. CMV - Cucumber Mosaic Virus

5. CTV - Citrus Tristeza Virus

6. DSIR - Department of Scientific and Industrial Research, New Zealand.

7.   ELISA - Enzyme Linked Immuno-Sorbent Assay

8. EU - European Union

9. FAO - Food and Agriculture Organisation of the United Nations

10. FAOHQ - FAO Headquarters

11. FFA - Forum Fisheries Agency

12.   IRETA - Institute for Research, Extension, and Training in Agriculture

13. MAFF - Ministry of Agriculture, Fisheries and Forests (Tonga).

14. MAFFF - Ministry of Agriculture, Fisheries, Forests and Food (new Tongan Ministry)

15. MSc - Masters of Science

16. NPPO - National Plant Protection Organisations

17. NPPS - National Plant Protection Organisations

18. PPPO - Pacific Plant Protection Organisation

19. RPPO - Regional Plant Protection Organisation

20. SPEC - South Pacific Bureau for Economic Co-operation

21. SPC - South Pacific Commission

22. SPC - Secretariat for the Pacific Community (new name for SPC)

23. USA - United States of America

24. USP - University of the South Pacific

25. UNDP - United Nations Development Programme

26. VNPV - Vanilla Necrosis Potyvirus

www.ingramcontent.com/pod-product-compliance
Lightning Source LLC
Chambersburg PA
CBHW060607200326
41521CB00007B/688